Which way is up?

And other stupid questions?

By

Damon Dion Reed

Table of Contents

Chapter 1: Well I'll be a Dinosaur's Uncle? 3

Chapter 2: Redefining Adiabatic Expansion 6

Chapter 3: Vroom-Vroom Space Plane! 11

Chapter 4: Batteries 16

Chapter 5: Intel Idea 22

Chapter 6: Alpha Particles 27

Chapter 7: Watch Glass 30

Chapter 8: Weak Nuclear Forces 35

Chapter 9: Atomic TLC 40

Chapter 10: Gravity and Tilt 51

Chapter 11: Splitting Centripetal Acceleration 58

Chapter 12: Propulsion & Fusion 61

Chapter 13: Termites 65

Chapter 14: Seeing & Communicating 67

Chapter 15: Which way is up? 70

Chapter 1: Well I'll be a Dinosaur's Uncle?

Here's an odd thought: Beaks came **before** teeth? (Yep, it's going to be one of those books: Call out the National Guard to protect the children from…from…SCIENCE?) I mean, if you think about it, it makes perfect evolutionary sense: First, there was single celled catalytic organisms; Next, there were shellfish; Then, there were things with bones; Then, there were things with bones sticking out of them, i.e. beaks; And finally, there were teeth?

I know that scientists have postulated that dinosaurs "might" have had feathers, but T-Rex does sort of look like a bird…with a major pituitary problems? And what might cause pituitary problems? Anybody? (FYI, Radioactive Iodine.) Or in simpler terms, on Moon-Earth, with the massive Volcano Zit spewing-out crap, AS WELL AS, countless asteroids knocking Mars, Venus, and Mercury into their planetary orbits, it's NOT outside the realm of consciousness that an asteroid with high levels of radioactive iodine disintegrated in the atmosphere, which caused a group of birds to grow into dinosaurs. Or in the simplest terms, just imagine Superman's spaceship being a radioactive-iodine-asteroid, crashing into a corn field, and some birds eating the radioactive corn, which caused them to grow into MASSIVE dinosaurs...with teeth. Any who, back to beaks.

In theory, if you start-off with a beak, then all you need is some skin-flaps to grow over the beak, which should allow the beak to partition into teeth. (If you can't imagine what skin flaps look like, just look at a turkey?) Actually, let me put that another way.

If beaks were the dominate "mouth" on Moon-Earth, then the massive diversification BEFORE the advent of dinosaurs would ensure that beaks remained after the dinosaurs, i.e. beaks still exist? (Or in bone terms, there are more examples of bones splitting apart to form two bones, than bones fusing together to form one bone?) And finally, the MOST damning evidence of beaks forming first is...what for it...just a little longer...ants have beaks! I mean, they're NOT called beaks, but they're exoskeleton mandibles used to munch on food. All of which, brings us to ants, the second largest biomass on Earth.

I know that humanity thinks that cannibalism is a horrific thing, butt...the second largest biomass on Earth, ANTS, are cannibalistic. I mean, ants will fucking eat anything...including other ants. So, before you go to the polls in 2020, ask yourself this: Has humanity evolved past ants? Or, are we worse than ants? Oh wait, we have God in our lives and that makes us so much MORE "heavenly" than ants?

In conclusion, which way is up? (FYI, GMOs are up. #Word) Or in other words, the time between "life events" is somewhat bothersome. For example, the time between birds and dinosaurs, i.e. beaks and teeth, is quite large. AND, if the evolution of ALL smaller animals with teeth were the result of the evolution of dinosaurs, then the dinosaurs probably existed a lot longer than humans, i.e. dinosaurs were our uncles. And finally, birds aren't dumb; I've seen birds scavenge for dead bugs off the grills of trucks...so they do have the capability to learn, which is more than I can say for some people? #TurkeySkinFlaps (That wasn't a great introduction because I didn't mention Quanta Dynamics?)

Chapter 2: Redefining Adiabatic Expansion

Here's a weird question: What's the evolutionary force for flight? Well, via the "old" Pangea theory, where the animals walked everywhere, there's no REAL evolutionary force for flight. Butt, with my Frosty Moon-Butt theory, where water was sparsely available over HUGE distances, NOW you've got an evolutionary force for flight. I mean, spores and seeds can float over HUGE distances via wind energy, butt "walking" animals require food. Or in other words, when Moon-Earth began to rotate, valley lakes of vegetation began to spring-up; BUTT, animal life couldn't spread-out because of the lack of vegetation between the valley lakes of life. Therefore, the evolutionary force, i.e. survival, favored the entity that could travel the farthest without needing much food or water. (Now bird migration makes a little more sense?) All of which, resulted in birds "carrying" smaller bugs and seeds around the world...until there was enough vegetation for animals to walk between the valley lakes of life.

If you've ever sat and looked at a fly, you'll realize it has dainty wings and a fat-ass. Conversely, airplanes have HUGE wings, in order to lift their smaller-asses into the air. In any event, as weird as this sounds, flying is like swimming. I mean, flying entities are simply doing the air breaststroke? In any event, highly isolated valley lakes of life with almost NO predators is the ONLY way to imagine chickens getting too fat to fly? (BTW, imagining chicken's great-great ancestors having to walk-up mountains as the Frosty Moon-Butt melted, is somewhat comical?)

Last, but not least, there's the question of lift as it relates to atmospheric concentrations. Or in other words, was it easier to fly when Moon-Earth had a HIGHER oxygen content? Now, in order to answer that question, we need to think about the essence of gravity as it relates to flight. Previously, I postulated that flight was the result of adiabatic gas expansion under the airplane wing, which resulted in the movement of negative plasma towards the adiabatically expanding gas, i.e. the "force" of lift. Or in simpler terms, it's like making the underside of the wing relatively positive, which causes negative plasma to move towards the bottom of the wing. Actually, let me try and explain that a little better.

The whole bases of flight is the rate of air movement about the wing. Now, the reason why this is important is because of the following: When you punt an atom's atomic orbitals, there's a rate associated with the atomic orbitals returning to their normal "protective" state around the atomic nucleus. Or in simpler terms, look at the next figure...PLEASE?

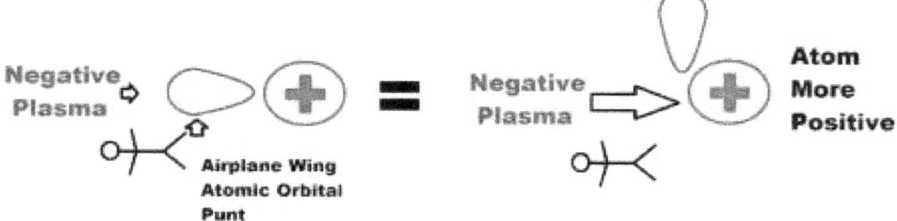

Figure 1: Airplane Wing Atomic Orbital Punt

As you can see, the airplane wing punts an atomic orbital, which allows the negative plasma to surge towards the atom. And as a result of the amount of time the atom spends under the wing, the more negative plasma, i.e. force, can surge towards the more positive atoms, i.e. lift. All of which, is a function of negative plasma diffusing away from the Earth's crust, towards deep-dark-cold POSITIVE space, i.e. Earth is NOT an energetically closed system.

Now, plasma isn't the ONLY culprit to adiabatic gas expansion. Quite simply, since "atomic orbital shells" order themselves based upon the underlying magnetism, it shouldn't be a surprise that the distortion of an atomic orbital will allow the Earth's MEQ to create a pseudo-plasma bond where the atomic orbital used to be, which will INREASE the time it takes for the atomic orbitals to return back to its protective state. Or in simpler terms, Earth's MEQ and plasma combo increases the time of adiabatic expansion, which allows for greater lift.

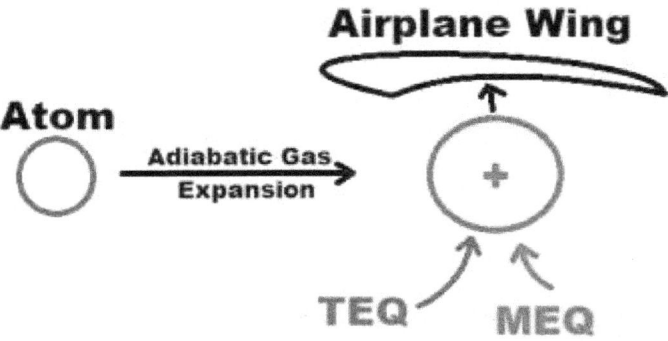

Figure 2: Improving Refrigerators with Magnets?

As you can see, by the title of figure 2, I just like asking questions? Also, "adiabatic gas expansion" is nothing more than the distortion of negative atomic orbitals, which increases the "expressed" positivity of the atom. All of which, causes relatively negative energy to surge towards the atom, i.e. the force of lift.

With all this in mind, it's a theoretical "fact" that atoms undergoing adiabatic gas expansion become MORE paramagnetic. (I've always wanted to use the term: Theoretical Fact. It just sounds funny. Also, when you distort PART of the atomic orbitals, the underlying magnetism, which aligns the atomic orbitals, escapes, i.e. paramagnetism?) All of which, brings us back to the fat-ass fly. But first, let's review electromagnetic induction...just to be annoying?

Now at first glance, adiabatic gas expansion and electromagnetic induction have ABSOLUTELY nothing in common. (Also, how does diamagnetism squeeze into all this?) But, when you put a fly's **protein** wing in-between these two concepts, it opens up the concept of biopolymers, which could make planes float higher in the sky. (I'm telling you'all, this biopolymer shit is a fucking gold mine.) All of which, will make a little more sense in the next chapter?

In conclusion, there are many factors in the evolution of flight. First and foremost, the high level of oxygen on Moon-Earth aided in the ease of flight, because the atomic orbitals of oxygen are easier to distort in O_2 in comparison to CO_2, i.e. greater possible lift in atmospheres with higher amounts of O_2. Second, and equally important, the higher level of O_2, which is electronegative, attracted the relatively positive MEQ on Moon-Earth, i.e. the magnetosphere was denser about layers of the atmosphere with higher amounts of electronegative O_2. And finally, the evolutionary force to flight is a function of water being spread-out between valleys lakes of life…on Moon-Earth. (BTW, it's possible that chicken feathers didn't provide as much lift in an atmosphere with more CO_2, which caused them to eat their emotions…because they were sad about not being able to fly anymore? #ForeShadowing)

Chapter 3: Vroom-Vroom Space Plane!

As we ALL know, physics can NOT explain how birds fly. Seriously, their "calculations" can NOT describe how birds fly. Granted, the calculations are based upon a "closed" system, where the Earth is **not** CONSTANTLY degrading and releasing plasma into space, but that's beside the point. The point is: Humanity didn't TRULY grasp the concept of magnetism until AFTER we discovered how to fly.

In previous books, I postulated that magnetism could be used to separate oxygen from nitrogen, which could be used to enhance the engine's thrust AS WELL AS allow the plane to fly higher. Any who, here's another idea.

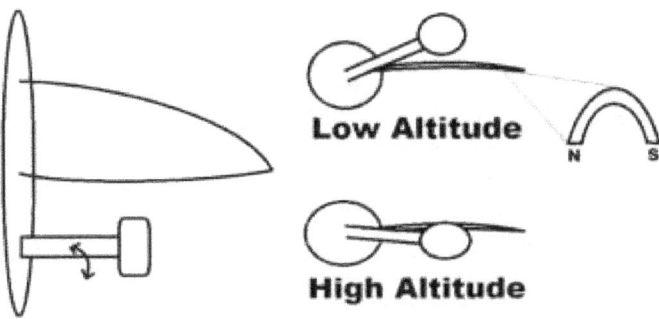

Figure 3: So Simple?

As you can surmise, thrust is dependent on airflow INTO the jet engine. Or in other words, the movement of the engines to below the wing will allow the concentration of more air INTO the jet engines, which will allow the airplane to fly higher. (FYI, this idea made the Air Force cream its shorts, which is why they let me move to Denver?) All of which, brings us back to magnetism...and biopolymers.

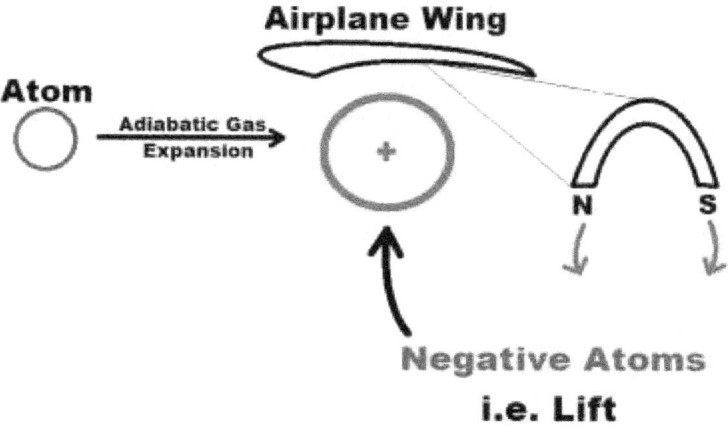

Figure 4: Just an Idea?

In theory, and this is important, magnetism under the airplane's wing will result in greater drag. But, this is a necessary visualization...as it relates to understanding the magnetism of biopolymers. Or in simpler terms, the enhanced magnetism under the wing will create pseudo-plasma bonds with the magnetism being released by non-valence atomic orbitals, which will improve the "adiabatic expansion" of the air under the wing. Or in other words, the magnetism under the wing is increasing the TIME the atom is "relatively" positive, which increases the amount of other negative atoms moving towards the "relatively" positive atoms, i.e. lift. All of which, brings us to my zeolite-magnetism idea.

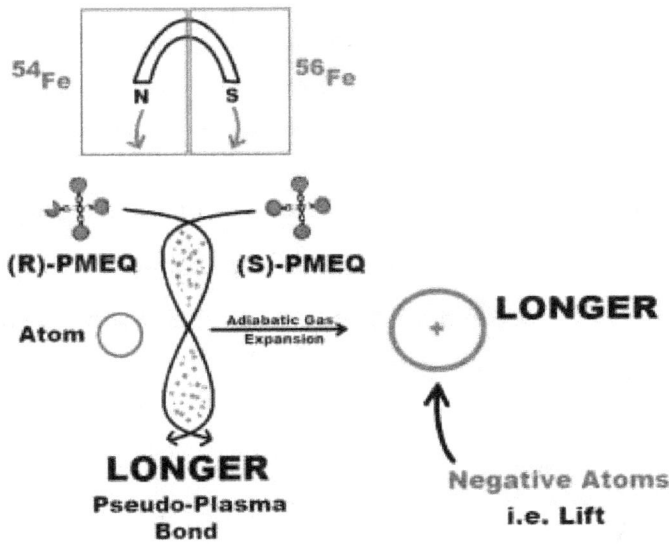

Figure 5: Hello Lift?

This is going to be super confusing...so hang in there? Quite simply, based upon the electronic interaction of the different Neuproz Clusters of iron 54 & 56, each isotope will emit a UNIQUE spectrum of PMEQ. And since the similarity between (R) and (S) PMEQ determines the relative rate of swirl decay, the magnetic swirl decay under the wing will be SLOWER, which will increase the availability of the PMEQ to distort the atoms under the wing, by making pseudo-plasma bonds with the elements. Or in dancing terms, there's less attraction between the dance partners released by iron 54 & 56, which means they'll switch partners and dance MORE with the atoms under the wing...to produce stronger adiabatic gas expansion. All of which, brings us to my idea on how to build these magnets.

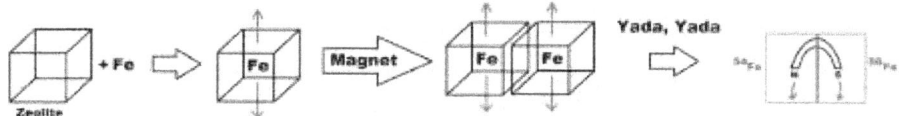

Figure 6: Yada, Yada?

As you can see, the "Yada, Yada" is the REALLY tough step. My suggestion, if you care, is to do small portions, then align them to make a bigger magnet. But enough about that, back to the point at hand: Biopolymers.

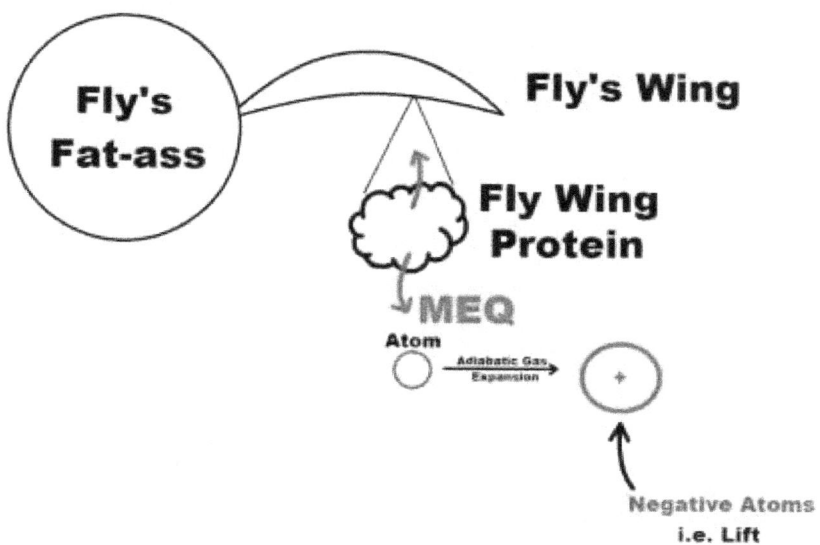

Figure 7: Biopolymers

For those of you who don't know this, proteins can have dipole moments, which means there's a "focusing" of MEQ within the protein.

And as a result of this, the MEQ can distort the atoms under the wing, which will cause the atoms undergo adiabatic gas expansion, i.e. an increase in their expressed positivity. All of which, creates lift. (Pretty amazing proteins…if you ask me?) All of which, is WHY current physics can NOT explain HOW flies lift their fat-asses into the air?

In conclusion, I'm not perfect, but I'm still trying to help. Specifically, underwing biopolymers and moveable engines should allow planes to fly even higher. And with the oxygen purification device, that I've previously postulated, an airplane should be able to collect enough liquid oxygen to create a vroom-vroom space plane? All of which, is based upon the postulates created by Quanta Dynamics? (As for the zeolite-magnetic stuff, that's going to make some amazingly efficient cars…and technology?)

Chapter 4: Batteries

In order to make isotopically pure elements, we need to understand the "process" by which the negatively diffusive stars facilitate the isotopic relaxation/degradation. All of which, should explain how and why isotopically pure elements will make superior components for technology. But first, a quick review of "mass."

We exist in a negative branch of the universe that is ordered based upon the electronic interaction between galaxies, stars, and planets. And even though we want to believe in the finite mass/volume of ALL matter, the mass/volume is relative to the environment and internal electronics. Therefore, since I already talked extensively about the environment factors, let me muse upon the inner electronics, which might explain why some data can only be reproduced in certain super-colliders.

For a moment, let's imagine that the entropic expansion/decay of a proton results in 6 unique layers of MEQ. (BTW, this would explain why protons are positive, i.e. MEQ are relatively positive?) And since protons are always decaying and releasing MEQ, each unique layer will release unique MEQ at unique rates. All of which, brings us to the super-collider "data."

Super-colliders use unique magnetic fields to focus protons, which is a problem because protons have a ton of MEQ. Or in simpler terms, the unique magnetism is uniquely distorting the unique protons, which results in UNIQUE data that's ONLY reproducible in THAT super-collider. Now, factor in the possibility of 6 unique proton layers and you get a HUGE quagmire. Or in highly educated terms, for the last thirty years, there has been an insipid argument between scientists, which is based upon WHICH super-collider from which they ascertain their data? (I get extra points for using "which" three times in the same sentence?) All of which, brings us to a very interesting point.

In theory, the stronger you make the super-collider's magnets, the faster you can accelerate the proton, but the FASTER the protons will degrade. Or in other words, there's a reverse correlation between acceleration and focusing-longevity. Or in Einstein terms, the increase in mass do to acceleration is the result of energy being destroyed...(cough)...I mean, the super magnetic environment is destroying the MEQ in the proton? Or in simpler terms, protons degrade faster in stronger magnetic environments, which is why weaker super-colliders can focus protons for longer. All of which, brings us back to the 6 unique layers of a proton.

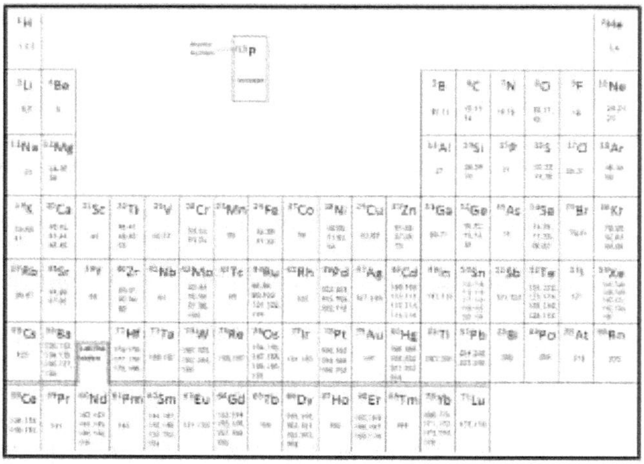

[Periodic table image with adjacent table:]

Base Neuproz Grouping	Layers
[N]	1
[3N]	2
[5N]	3
[7N]	4
[9N]	5
[11N]	6

Figure 8: Hey look, a figure?

If a proton expresses 6 distinctive electronic layers in the environment around our Sun, then it should be logical that heavier elements, with more protective electrons, can exchange "stronger" negative Neuprotrons via different layers. All of which, not only gives a method to the stronger electronics of larger Neuproz Groupings, but a REASON why our bodies are designed to continuously flush elements through it. Actually, let me try to explain that better.

If you remember in my last book, *I Am Prism…And so can you?*, I talked about negative plasma being trapped within the Neuproz Cluster, which facilitates Isotopic Relaxation.

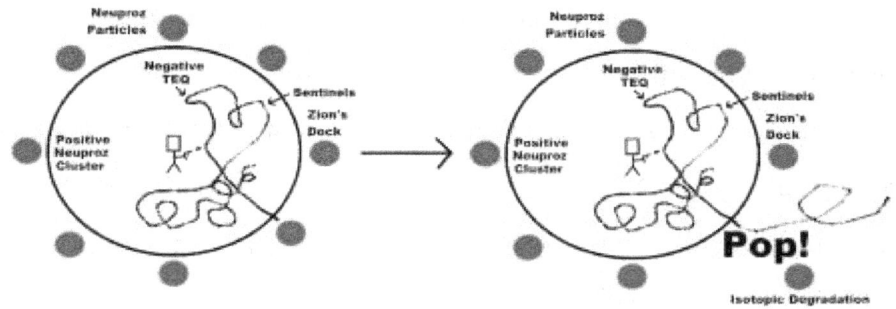

Figure 9: The Matrix Loading Dock. (From: *I Am Prism*)

Unfortunately, since elements last SOOOO long, this "process" CAN occur slowly, which causes the element to have slightly different characteristics. For example, if Sulfur has a half-life of a billion years, then it's conceivable that this degradation process could take up to five years in the right environment, which MEANS five years of weird transitional elemental characteristics. All of which, the human body might not like if the transitional elemental characteristics distorts a catalytic pathway.

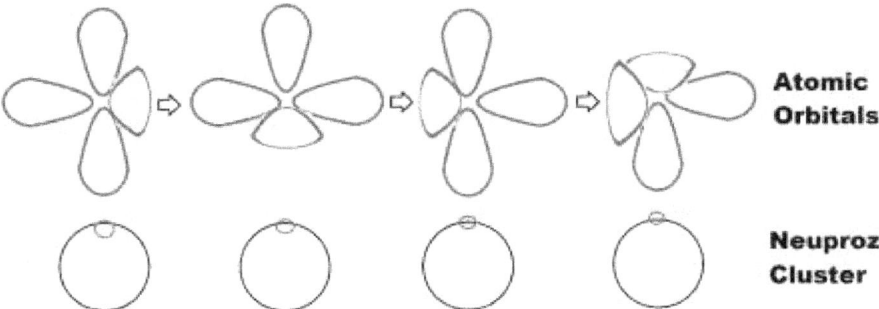

Figure 10: SLOW Isotopic Relaxation/Decay

As you can surmise, since the body is based upon catalysis, which is based upon CERTAIN atomic orbital characteristics, then ANY distortion of these atomic orbital characteristics will inhibit a catalytic pathway. Or in simpler terms, the rate of cellular division is a function of Isotopic Relaxation/Degradation? All of which, brings us back to superior technology of isotopically pure elements.

Even though I've ONLY chit-chatted about the degradation of protons, you have to remember that electrons are also continuously degrading, which complicates the Isotopic Relaxation process. (FYI, someday, someone will notice the evolution of isotopic purities in recycled electronics?) Any who, I have no clue as to how many layers an electron has, but it seems logical that the slow degradation of electrons DOES occur. And since electrons traverse the Neuproz Cluster to recharge via sating the positive charge therein, it is abundantly logical that LESS negative electrons will cause more electronic repulsion in the Neuproz Cluster, which will facilitate Figure 10. Or in simpler terms, the degradation of electrons speeds-up the process of Isotopic Relaxation/Degradation, which brings us to your phone battery.

Since your phone battery is constructed based upon the movement of electrons through nano-structures, the distortion of the atomic orbitals in these nano-structures will DRASTICALLY effect the batteries ability to charge and recharge. Or in Quanta Dynamic terms, heavier isotopes are more sensitive to "degraded" electrons, because they have greater charge density in the Neuproz Cluster. Or in engineering terms, decayed electrons will cause the atomic orbitals of the heavier isotopes to SAG, which will cause cracks in the battery foundation. (BTW, if someone derives an equation that accurately determines the battery life-time based upon the Isotopic Purity of the components, then that might be worth a pretty penny?)

In conclusion, since atoms last for a long time, Isotopic Relaxation/Degradation ALSO lasts a long time…depending on the environment. But more importantly, the heavier isotopes are more sensitive to electron decays, which causes their atomic orbitals to shift in certain environments. Therefore, isotopically PURE batteries will last longer. And finally, this out of place thought: When a partially degraded proton is cast back out into deep-dark-cold positive space, the retraction of these partially degraded 6-electronic layers ONTO the core of the proton will distort the inner electronics faster, i.e. decrease the half-life of the proton? All of which, facilitates the fission of smaller elements to forge heavier elements?

Chapter 5: Intel Idea

Hopefully, after the last chapter, you're a little more open-minded to the idea of Isotopically Pure technology? All of which, brings us to the ultimate question: How do WE make Isotopically Pure elements? Well, let's start at a relative beginning.

Newton was a nerd's nerd and he made great strides in science. But, a lot has happened since his time, i.e. Anisotropy. Or in simpler terms, Quanta Dynamics. Or in more explicit terms, protons are always degrading and releasing Magnetic Energetic Quanta. All of which, brings us to the HEAD of the problem: Everyone is STILL trying to make isotopes as if protons are billiard balls, i.e. Newton's idea. Or in other words, the only conceivable thing you can do with "billiard ball" technology is create a pseudo-isotope, aka inclusionary-isotope. (FYI, the extra proton's random release of MEQ next to the atomic nucleus is probably the thing that fucks-up all the "nuclear forces?"

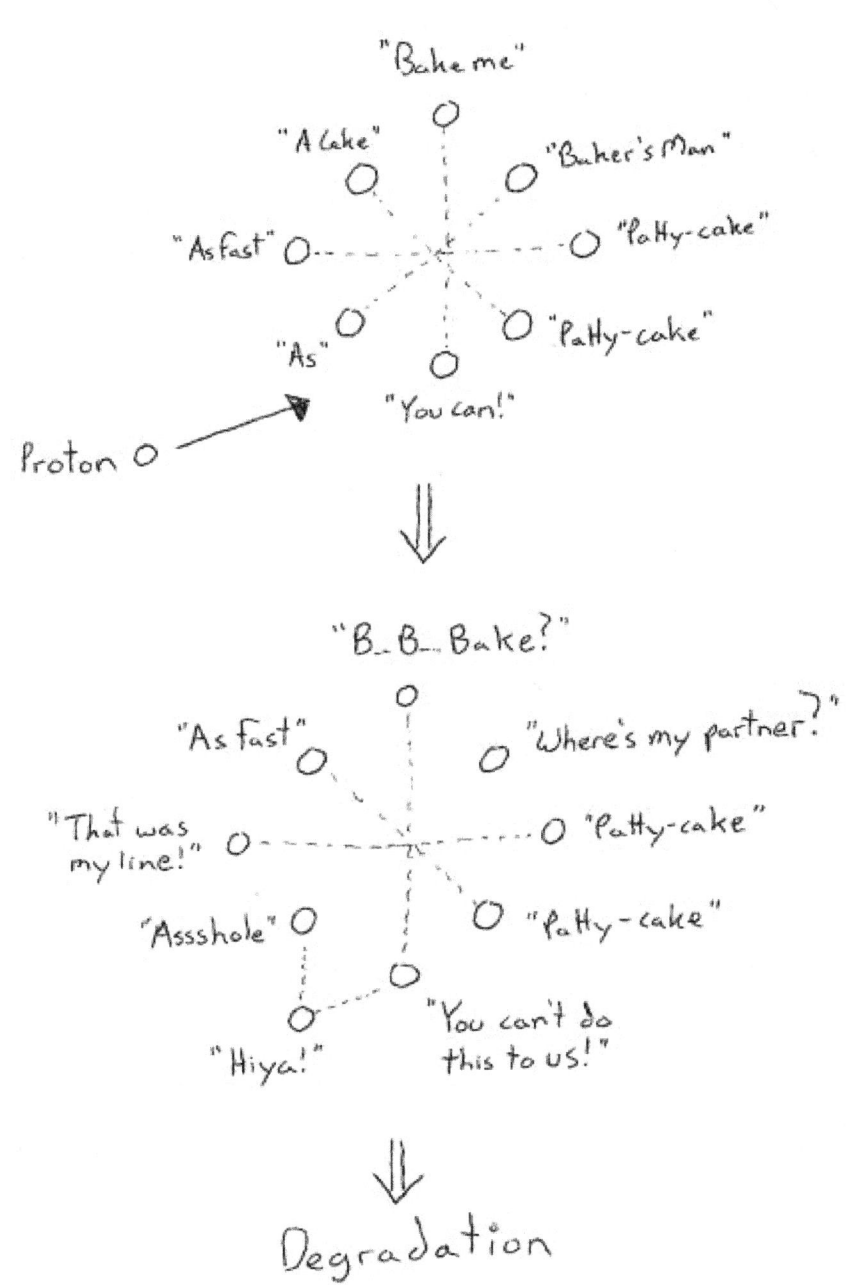

Figure 11: Childish Figure? (From *Quanta Dynamics*)

Granted, examples like this PROBABLY didn't help people believe in my theories, but that's beside the point. The point is, there are "nuclear forces" between the protons and neutrons that are SPHERICALLY orientated to form an atomic nucleus. Or in simpler terms, randomized proton and neutron placement MEANS there has to be random "nuclear forces" between these particles? All of which means, we NEED to stimulate heavier isotopes to Isotopically Relax/Degrade, to form STABLE isotopically pure technology. (I'll talk about the theory of this "electronic distortion" in subsequent chapters?) Any who, here's my idea on how to make Isotopically Pure elements.

We ALL know that microwaves degrade in the atomic orbitals of oxygen to produce Thermal Energetic Quanta…right? I mean, the other option is that the microwave reverberation of the milieu causes the distortion of the atomic orbitals, which causes the electrons to degrade and release TEQ? Any who, for whatever reason, oxides respond to microwaves via releasing TEQ. Therefore, since TEQ are negative and ALWAYS SURGING towards the positive atomic nucleus, where they cause heavier elements to undergo Isotopic Relaxation/Degradation, all we have to do is tweak the atomic orbitals…so **more** TEQ surge towards the atomic nucleus. (This is the hard part.) As for the rest, it's pretty straight forward.

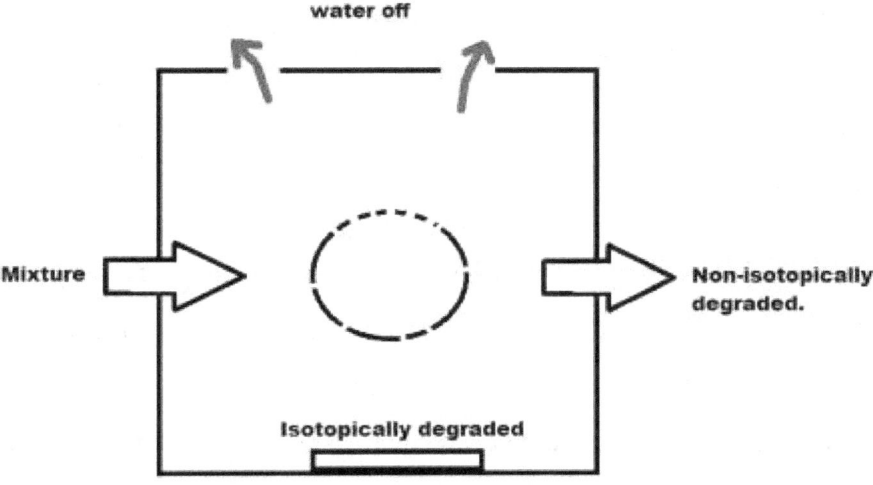

Figure 12: Microwave Reaction Chamber

Quite simply, the Isotopic Relaxation/Degradation of Silicon will release a neutron, which will quickly degrade into a proton. And since protons are positive and oxygen is negative, bada-boom, bada-bing. Or in scientific terms, the eventual redox reaction will produce water, which responds to microwave radiation and defies gravity, and molecular silicon, which does not respond to microwave irradiation, i.e. collects at the bottom of the reaction vessel. As for the other silicon isotope, the non-isotopically degraded oxide, you can just suck it off. (Sorry, I forgot to mention that the circle is a reaction chamber created by ONLY microwaves.) All of which, brings us the CATALYSIS of this reaction, i.e. the thing that HELPS distort the atomic orbitals so more negative TEQ surge towards the positive atomic nucleus.

My first choice of a catalysis would be light, because photons move faster than negative TEQ, i.e. you can slip photons through smaller cracks between the atomic orbitals. But then, I remembered that iron reacts violently with microwaves. So maybe both?

In conclusion, the purification of molecular silicon away from silica oxide and water is quite simple when using a microwave reaction chamber. The key to the whole thing is finding the "catalytic conditions," which probably involves some sort of knowledge ascertained SINCE the time of Newton? Or, we could just keep doing things based upon Newton's postulates?

Chapter 6: Alpha Particles

Another problem with Newton's "billiard ball" theory is that it completely overlooks acid-base chemistry. Or in scientific terms, what's the pKa of Helium? Hopefully, I've FINALLY got the attention of some nerds? Any who, let me try and explain this better?

If you know anything about radioactive decay, then you know that radioactive elements CAUSE other elements to become radioactive. As per the METHOD of this stimulus, well...that's the problem. You see, with the billiard ball theory, scientists believe that a fast moving alpha particle, i.e. helium atomic nucleus, shoots out of one atom, past the atomic orbitals of another atom, and crashes into the atomic nucleus of the adjacent atom, which is the first problem! Why is it a problem? Well, atoms are MOSTLY negative atomic orbitals and helium atomic nuclei are positive, which means the following: If an alpha particle does NOT collide with the MOST ADJACENT atom, then its velocity will be GREATLY reduced. Or in simpler terms, from the moment an alpha particle is released, it is continually slowing down, which means it's less likely to make it past more distant atomic orbitals.

So let's recap: 1) Radioactive elements have a HIGH efficacy of making adjacent elements radioactive; 2) Alpha particles are positive and slow down quickly in this negative branch of the universe; 3) Atomic nuclei are super small and difficult to aim for. All of which means, the HIGH efficacy of a radioactive element creating adjacent radioactive elements is the result of acid-base chemistry. Or in simpler terms, look down.

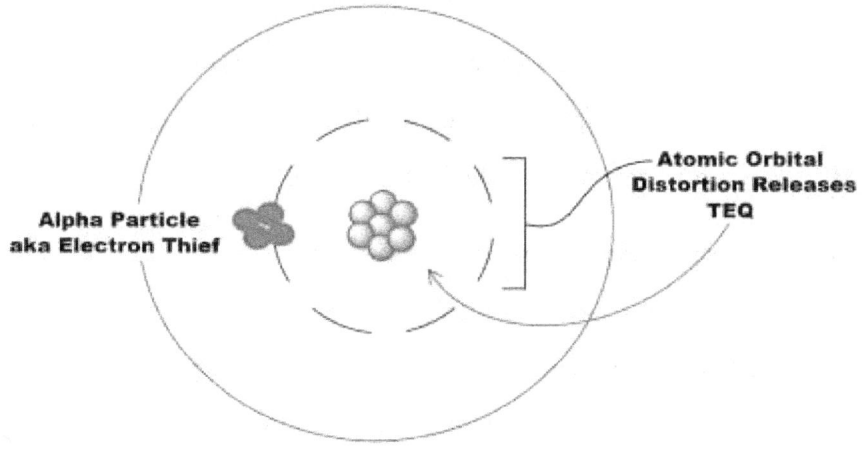

Figure 13: Hello Mr. Logic?

As you can see, the alpha particle STEALS some non-valence electrons, which causes extreme TEQ release via atomic orbital distortions. And, as we all know now, TEQ near the atomic nucleus causes Isotopic Relaxation/Degradation, aka radioactive decay. (FYI, please don't forget that TEQ inside the Neuproz Cluster ALSO causes the degradation of negative Neuprotrons, which is another factor in Isotopic Relaxation/Degradation.) Sooooo SCIENTISTS, does that make MORE sense than the flawed billiard ball explanation?

In conclusion, Newton's work has undoubtedly pushed science forward. But, nobody is perfect. And as such, I'm pleading with the current scientists to take another look at the logic behind the "billiard ball theory," as it relates to Quanta Dynamics. And finally, please take a deep-breath before starting the next chapter; it's a theoretical doozie?

Chapter 7: Watch Glass

So here is the logic: If you COMPARE galaxies to stars to planets to Neuproz Clusters to protons & neutrons, then you'll see a SIMILIAR watch glass posture to the "energy." And before I get to all that, I'm going to use pyramids as a "visualization" technique to explain the movement of energy. So...why can't we teach this pattern of energy in science? What does this pattern mean: There is a God and what we do with our time matters? (Maybe?) We can do better? Any who, Quanta Dynamics.

Scientists already have names for the forces between the particles in the atomic nucleus, but because of what was explained in chapter 4, nobody can agree on the "ENERGY" being exchanged by these particles. Or in simpler terms, the data "created" by the super-conductors is what drives the postulates with regards to the weak and strong nuclear forces. Therefore, for the course of this chapter, let's ALL start with the premise that there are "strong" and "weak" nuclear forces, THEN go from there? Or in other words, what do researchers know?

Well, they "know" that they can make short-lived radioisotopes, which they can use to SCIENTFICALLY help other people. The question is, which I don't have an answer for, is as follows: What is the statistics surrounding the half-lives of the elements created in reference to the half-lives of the naturally occurring elements? (That's a tough scientific question.) For starters, what do we believe in? Do we believe in God, Science, or both? Quite simply, the statistics surrounding the half-lives question is based-up your understanding of infinite, which is a complex and misleading word. So there you have it: The reason WHY scientists don't like my theories? Or, why Religious scholars fear science? (I don't know anymore?) In any event, let's move back to what is agreed-upon: Strong and weak nuclear forces.

If there are weak nuclear forces that are surging around the atomic nucleus, then in what sort of pattern would they be moving? A spherical pattern? Also, if there were NO lanes of traffic for these weak nuclear forces, then would it be MORE **efficient**? (Of course not.) Therefore, and I say this hesitantly, let's assume that the energy is moving around the atomic nucleus in rings, i.e. lanes of energy. And finally, based upon the extrapolated "ring-esk" structure of galaxies, stars, planets, and atomic nuclei, there is a "variable" fat center. All of which means: MORE energy is surging around the center of the atomic nucleus in comparison to the tops as bottoms, with regards to the strong or weak nuclear forces? Any who, this would mean that there are tiny watch glass proton/neutrons arrangements that FOCUSES this energy in a "ring-esk" highway? And now, my "pyramids visualization" technique to explain the movement of energy.

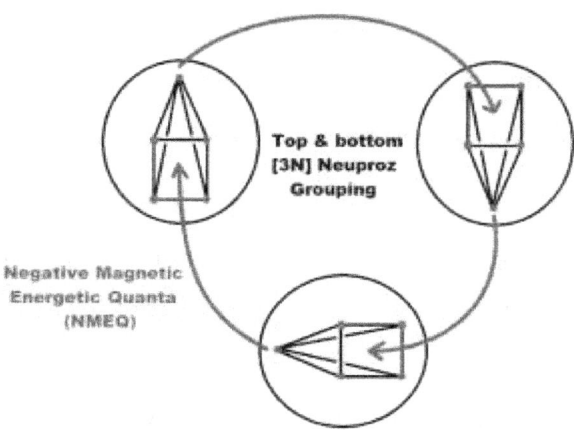

Figure 14: Three aligned atomic particles.

As you can see, my visualization technique, to show the direction of the NMEQ movement, should explain the "ring-esk" energy arrangement in protons and neutrons as it relates to the energy arrangement in galaxies, stars, planets, and atomic nuclei. Actually, on second thought, maybe I should draw if differently.

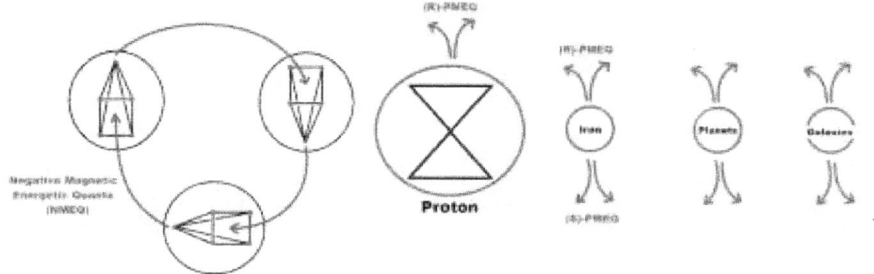

Figure 15: Extrapolation?

Hopefully, and I mean it, you will notice the TOP or BOTTOM of a Neuproz Cluster, aka Tritium, has BLUE NMEQ and the proton to the right, has RED PMEQ drawn coming out the top, which is completely confusing until I say "complexity?" (Totally lame reasoning, I know, but...?) Or in simpler terms, the MOVEMENT of the NMEQ between the protons and neutrons SLOWS the decay of these energetic quanta, which results in LESS red PMEQ being released. All of which, brings us to this question: As previously mentioned, why does the ANGLE between these atomic particles seem to "matter" as it relates to the AMOUNT of energy contained in these rings? And now, atomic spin?

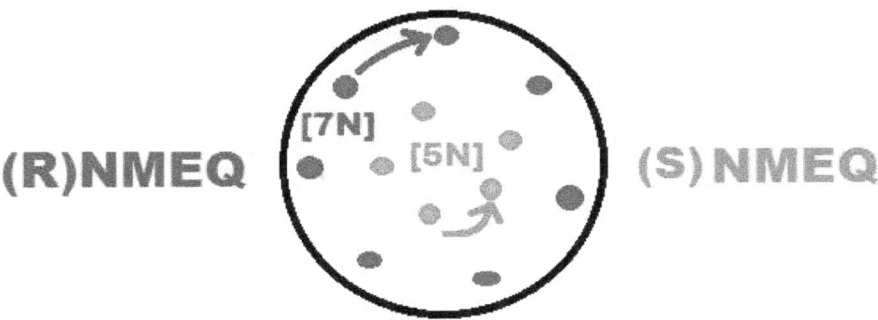

Figure 16: Atomic Spin

As you can surmise, the movement of this energy around the TOP, or BOTTOM, of this Neuproz Cluster is opposite in SPIN. And as I mentioned in the last book, *I Am Prism – And so can you?*, opposite spin MEQ degrade faster. (As for the precise reason why, it probably has to do with the precise overlap of opposite energy?) Anyway, here is my postulate on how and why different Neuproz Clusters have different spins based upon the Angular Distortion of these Neuproz Rings by extra NEUTRONS.

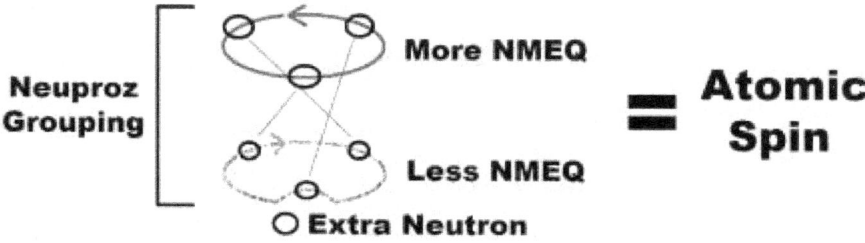

Figure 17: Angular Distortion

As you can surmise, the "weak" nuclear force, between the Neuproz Particles that are exchanging NMEQ, is the NUMBER of NMEQ. Or in simpler terms, the **NUMBER** of NMEQ surging around the TOP of the Neuproz Grouping is more than the BOTTOM of the Neuproz Grouping, as a result of the angular distortion caused by the extra proton, which results in ATOMIC SPIN?

In conclusion, I think we can extrapolate the shape of MOST of the proton's energy from the shape of galaxies, stars, and planets. And, since atomic nuclei are nearly perfect batteries, I think it is safe to assume that there is some ENERGY that's exchanging within the atomic nucleus that SLOWS the degradation of the energy therein, i.e. the sharing of (R&S) blue NMEQ about the TOP and BOTTOM of the Neuproz Groupings, which prevents the degradative release of (R&S) red PMEQ. And finally, the Angular Distortion caused by extra neutrons, i.e. the inhibition of the exchange of energy, with regards to the TOP and BOTTOM, causes some atomic nuclei to have SPIN.

Chapter 8: Weak Nuclear Forces

In as much as the variable "charge" of negative Neuprotrons is extremely interesting, the movement of opposite spin (R&S) NMEQ about mid-plane of the Neuproz Cluster MAYBE more interesting...and possibly, "stronger?" (Hold for inappropriate nerdie gasps.) But, before I get to all that, and some other nerdie stuff, let me try and paint you a picture of how I see the "forces" about [5N] Neuproz Grouping.

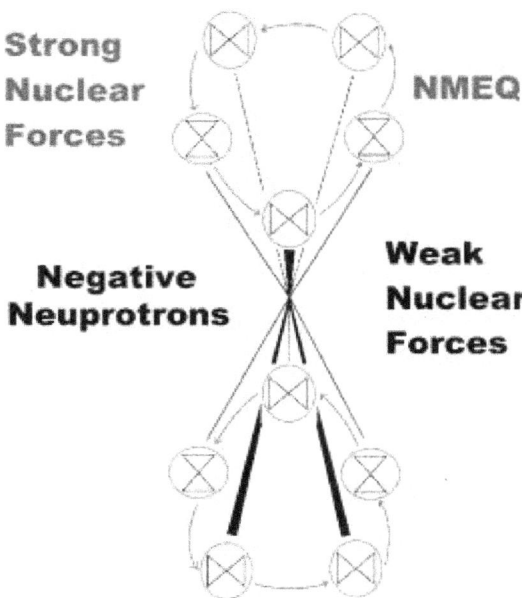

Figure 18: Watch Glass Forces

As you can see, there is a negative Neuprotron exchanging between each Neuproz Grouping. Also, there are (R&S) blue NMEQ surging around the particles in the TOP and BOTTOM of the Neuproz Grouping, i.e. the ring-esk super-highway of NMEQ. All of which, brings us to the next theory.

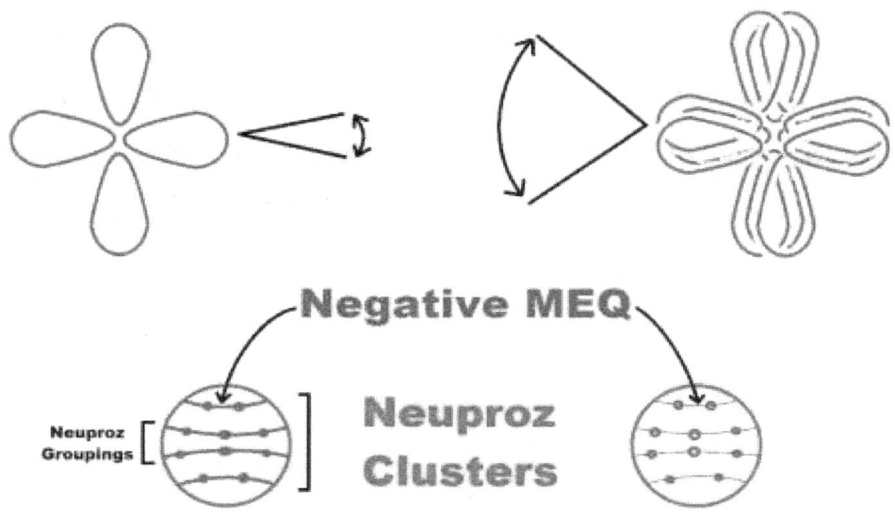

Figure 19: Electron EXPRESSION Factors.

In theory, there's probably a NUMERICAL spectrum of NMEQ that can surge around these ring-esk super highways, which results in the variable expression of the atomic orbitals, i.e. the wobble of the atomic orbitals in reference to the constant environment. But more importantly, this suggests the following:

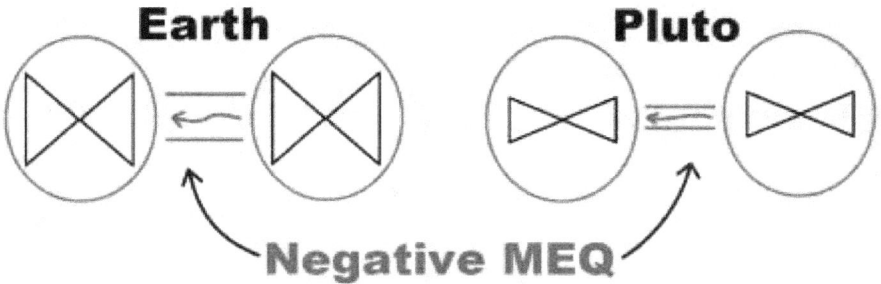

Figure 20: Uniqueness

Now at first glance, this should be confusing. First, Pluto has a less density than Earth. Second, if Pluto's matter has LESS NMEQ exchanging about the ring-esk super highways, then there should be greater positivity expressed about the atomic nucleus, which should explain the LESS density? (Actually, that seems pretty logical?) All of which, brings us to MASS.

Since Pluto's matter has LESS NMEQ surging about the atomic nucleus, then it is MORE positive. And if the matter is MORE positive, then it will absorb MORE of the Sun's negativity. All of which means, and this is kind of important, Pluto is a proto-planet, not a past-planet. Or in other terms, based upon density, when Pluto warms-up, it could become a Gas Giant, which is kind of troublesome if you think about it? In any event, back to the concept of nuclear forces.

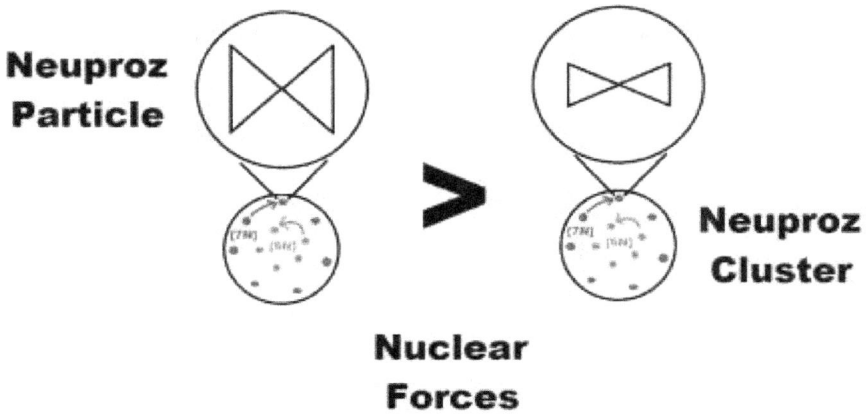

Figure 21: Nuclear Forces Spectrum

In as much as there's probably a spectrum of (R&S) NMEQ and Negative Neuprotrons, based upon the variable Watch Glass size, i.e. the NUMBER of (R&S) NMEQ surging around the ring-esk super highways, MORE THAN LIKELY, the "strong" nuclear forces are the ones created by (R&S) NMEQ?

In conclusion, hopefully I made a good argument as to HOW different types of matter can expand to become Gas Giants? Or in simpler terms, the smaller Watch Glass electronic structure of some matter increases the positivity of the atomic nucleus, which allows that matter to SUCK-UP more negative energy from a star. And since we all "agree" that negative energy, from stars, causes matter to degrade, it should be obvious HOW the degradation of this matter releases GAS…instead of just protons, which results in a TON of water? Or in simpler terms, the Sun's negative energy will cause Pluto's matter to release more alpha particles, i.e. Gas Giant, which is the result of inner electronics created by the Watch Glass structure. And finally, hopefully, I've made a good argument as to how the strong nuclear force is created by the (R&S) NMEQ surging around the circumference of the atomic nucleus? (Not the negative Neuprotrons surging about the middle of the Neuproz Cluster.

Chapter 9: Atomic TLC

As a consequence of the last chapter, someone might want to find a way to "measure" the difference between protons from Pluto and Earth. Fortunately, I have two ideas on the matter. (Pun or not, I still have two ideas on the topic?) Unfortunately, before I get to those two ideas, I need to talk about proton-NMR stuff...again.

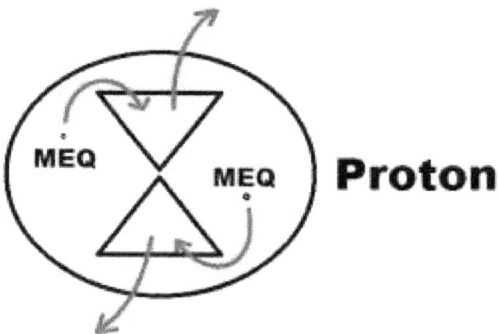

Figure 22: The root of Anisotropy?

As you can see, I'm postulating that protons are arranged like tiny atomic nuclei with degrading PMEQ layers, which traverse the center of the proton. As for the reason why I'm postulating this: Anisotropy DOGMENTS that protons directionally release MEQ...otherwise there would be NO anisotropy. Actually, let me try and explain that a little better.

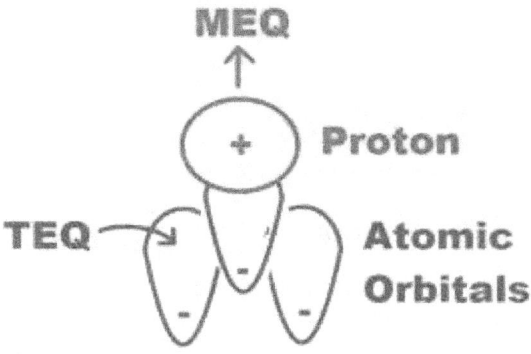

Figure 23: Star Ship Proton Enterprise

As a result of existing in a negative branch of the universe, negative TEQ are trying to surge towards the positive atomic nucleus, but they're collected by the atomic orbitals, which push the TEQ towards the positive proton. All of which, results in a Bed of TEQ, which causes the proton to degrade.

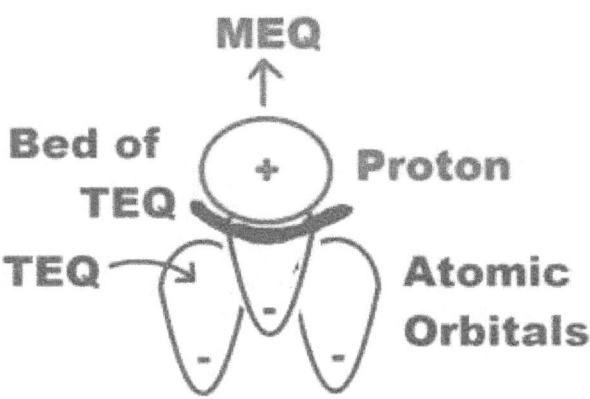

Figure 24: Bed of TEQ

Now, please take a moment and LOOK at what this picture is implying. Quite simply, if there was NO order to the proton, then the "bed of TEQ" would decay the bottom of the proton, which would cause the bottom of the proton to shoot relatively positive MEQ at the atomic nucleus, which would be bad. Why would it be bad? Well, negative TEQ will follow relatively positive MEQ in whichever direction they go. Or in simpler terms, if positive protons shot their relatively positive MEQ at the atomic nucleus, then this would decrease the half-life of the atomic nucleus by increasing the negative **flowrate** towards the positive atomic nucleus, aka Neuproz Cluster. Or in scientific terms, for some "logical" reason, protons SHOOT their relatively positive MEQ away from the atoms they are surrounding, which suggests a METHOD to the decay of the protons.

Figure 25: Protons PROTECT electronegative atoms?

Hopefully, you can surmise that the bed of negative TEQ causes the proton to degrade, i.e. dislodges MEQ, which follow the highly ordered path of energy, e.g. the Watch Glass, to shoot out the other side of the proton. (It's damn near analogous to the Neuproz Cluster, except with much smaller weird quanta of energy.) Or in simpler terms, the directional release of relatively positive MEQ sucks the relatively negative TEQ **away** from the element. All of which, brings us to the pinnacle of nerdieness.

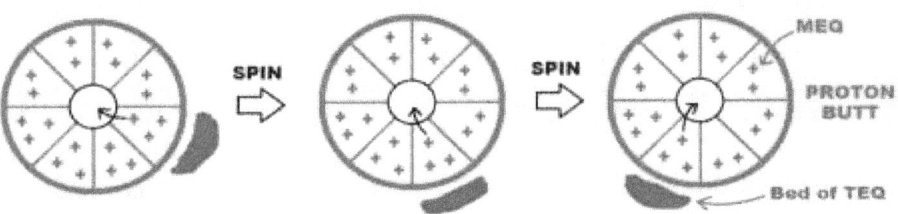

Figure 26: PROCESSION!!!

Hopefully, you can imagine HOW the decay of the proton, i.e. the movement of the MEQ through the Watch Glass and out the top of the proton, changes the charge within a certain quadrant of the proton, which causes the proton to shift its butt accordingly. Or in simpler terms, procession! Or in other words, the slow decay of the proton in association with the amount of negativity collected by the atomic orbital, determines the proton's spin rate. (I always wondered how and why procession occurred so beautifully...I mean...with ALL the randomness in the universe?) In any event, back to the problem at hand: My TWO ideas for measuring the difference between the protons on Earth and Pluto, i.e. measuring their Watch Glasses.

First Idea: The ISS Idea

As of right NOW, we can't detect the variation in PROTON masses, because the super-magnetic environments we use to DETECT mass SQUISHES all the matter to the same size. Or in other terms, the relatively positive MEQ used in super-colliders "squishes" the POSITIVE protons to about the same size. Or in totally nerdy terms, this is why certain groups of scientists rally around the data provided by DIFFERENT particle accelerators. Actually, let me find an outlet for my computer, so I can talk about this more.

Scientists get their "power" from their data and their data comes from the collision of particles. Unfortunately, different particles accelerators have different levels, and SPECTRUMS, of Magnetic Energetic Quanta. But more importantly, different super colliders use DIFFERENT protons. Therefore, we NEED a method to distinguish different protons, which are continually degrading.

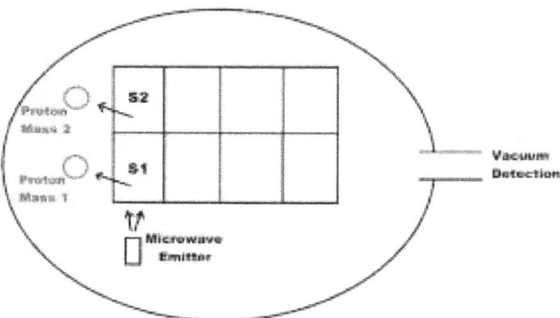

Figure 27: Intel-esk Idea.

We're on the precipice of something beyond profound…and yet, every scientist is screaming: You can't detect mass without a magnetic field, to discern the mass of the objects. All of which, brings me to my next idea.

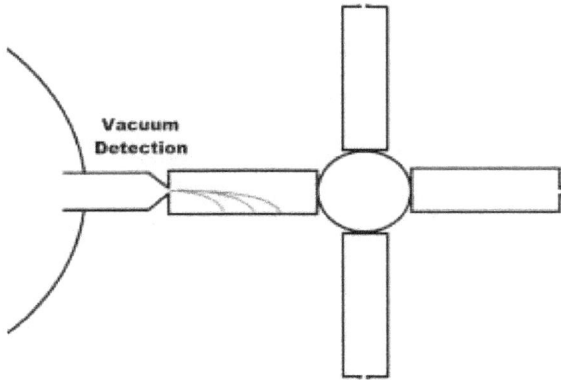

Figure 28: International Space Station-esk Detection

Hopefully, this looks like the INTERNATIONAL space station. Butt, for those who like to wonder past what already exists, this is something even more hideous: A non-magnetic mass detection apparatus. (FYI, the physicists are shitting a brick with questions like this: HOW the hell are you supposed to detect protons with this stupid drawing?) Well, how about this?

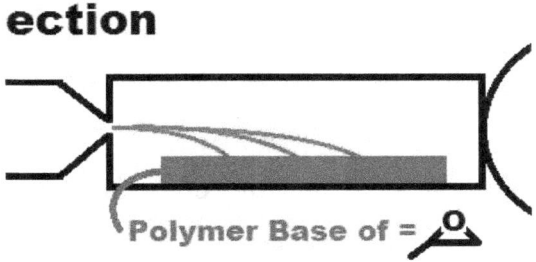

Figure 29: Polymer Detection

As you can surmise, based upon the International Space Station reference, God gave us a home with a specific magnetic field. And since one of the major factors in gravity is the diffusion of negative TEQ away from the Earth's surface, protons with different masses will travel different distances. Actually, let me expand upon this a little bit.

Gravity, as I understand it, is a complex thing that exists as an interwoven energy-display based upon charge, spin, and energy diffusion, which determines the relative rate of energy decay. For example, the relativity of directional release between (R)-MEQ and (S)-MEQ determines the rate of decay, i.e. how long it takes them to complete their swirl-dance-O-death. (See my last book, *I Am Prism*, for a longer discussion.) Any rate, #pun, mass is a function of expression as well as decay. Or in simpler terms, mass is fundamentally an expression of decay, i.e. the release of MEQ. Therefore, in order to "calculate" a more accurate mass of a proton, you need to eliminate the repulsion between the proton's MEQ and the Earth's MEQ, i.e. a Banana Radiation Shield OR Deep-dark-cold positive space. (Granted, we don't have the technology to escape the Sun's magnetosphere yet, but any "place" away from Earth will give a better measurement of the proton. Or in the simplest terms, the movement of the POSITIVE proton towards the NEGATIVE heat energy irradiating from the polymer base SHOULD be enough to differentiate between protons expressing different layers. (To separate protons within the same layer, you'll need to have an environment completely free of external MEQ?) Any who, back to the other specifics of this device.

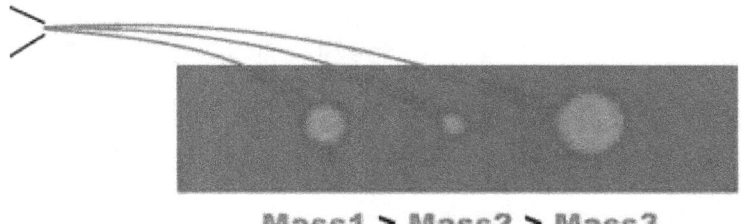

Figure 30: Polymer Spotting

As you can surmise, the protons that make the first polymer spot will have a greater positive mass than the other two polymer spots. All of which, leads to the following question: What does all this mean? Well, if you plate the PURE isotopes from different regions of the Earth onto Figure 17 in the last chapter, then you'll more than likely be able to tell WHICH pure isotope is "older," i.e. experienced a harsher environment. All of which, will totally take Carbon-14 dating to the next level? (BTW, I realize that this is based upon the decay of the same pure isotopes, which should only happen about similar degradation within the Neuproz Cluster, butt…in theory, based upon the directional orientation of the "stimulus" of the decay, i.e. from the bottom, each pure isotope will give two spots: A proton released from the TOP of the atom; A proton released from the BOTTOM of the atom, which will have experienced more decay, i.e. have smaller mass.) All of which, brings me to one last point.

In theory, higher energy super-colliders contain more MEQ, which should make the protons degrade faster, which has probably confused the heck out of the scientists. I mean, to a scientist, it would be utterly confusing as to WHY a super-collider with a stronger magnetic field can NOT keep a proton FOCUSED for longer than a "weaker" super collider, which has a "weaker" magnetic field...UNLESS...they understand Quanta Dynamics? Or in simpler terms, the STRONGER magnetic field is causing the protons to degrade faster in STRONGER super-colliders? (Trust me when I say that it is weird questions like these that move science forward...especially when nobody wants Science to move at all?)

Second Idea: The Better Idea

For those of you who don't know this, each planet has a different cone of magnetism, which tilts it differently towards the Sun. Therefore, if the solar system can separate planets based upon magnetism, and charge, then we should be able to do the same with protons. Now, the trick here is: ISOLATION. In theory, magnetic manipulation combined with a bit of Tender Loving Care, i.e. Thin Layer Chromatography, should be enough to separate protons with different cones.

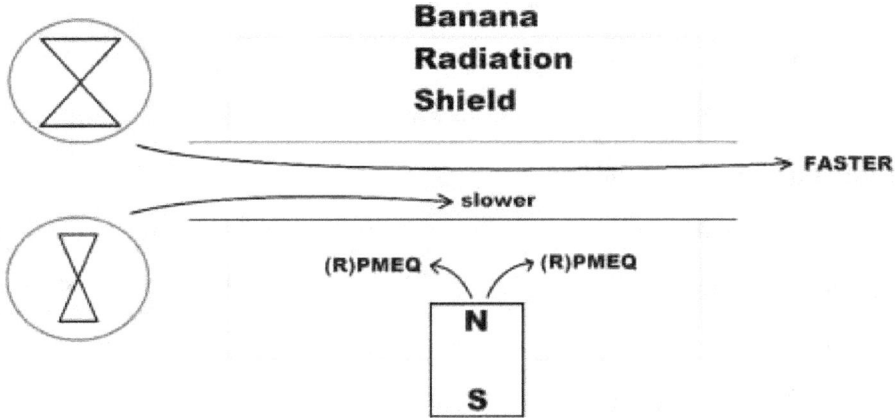

Figure 31: Atomic TLC

Hopefully, all you see is a cock and balls! (I think science is funnier with dick jokes?) Also, upon further examination of the figure, if you mind can handle it, you'll reach the following question: Why?

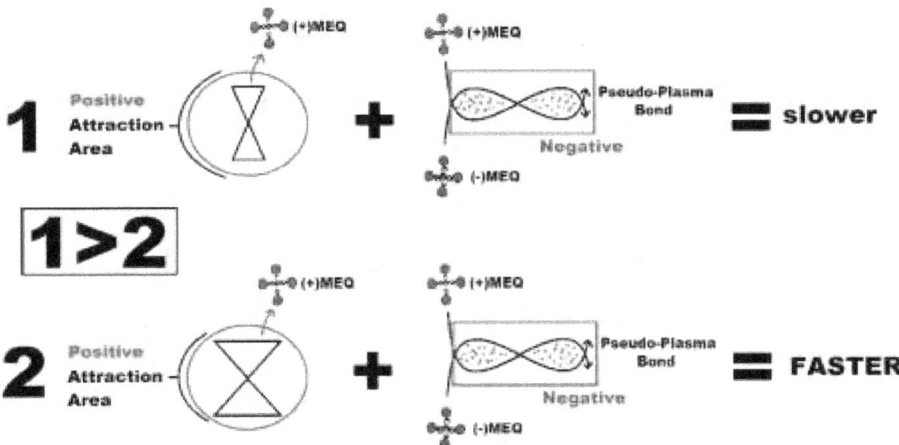

Figure 32: Positive Attraction Area

As you can surmise, the Positive Attraction Area determines the rate when all other things are held constant, i.e. the external magnetism.

Granted, the NARROWER the MEQ spectrum within with the Banana Radiation Shield will result in the BETTER proton separation, but that's beside the point. (The whole argument of HOW a spectrum of MEQ is produced in protons will be addressed later…hopefully.) The point is, now we have a way to purify unique protons. (FYI, the structural specificity of the Watch Glass probably decreases over the LIFE of a proton, which means fusion will only work with similarly structured Watch Glasses, i.e. protons of the same age.)

In conclusion, in terms of Quanta Dynamics, procession is a function of the atomic orbital's ability to corral negative TEQ. Or in terms current NMR books, procession is a unique function of unique atomic orbitals, which allows NMRs to accurately detect unique proton environments? Or in simpler terms, Anisotropy, the exchange of spin-information between two protons, would not exist if protons were releasing their MEQ towards the atomic nuclei. And as such, based upon protons being stimulated to decay by a bed of TEQs, i.e. the Star Ship Proton Enterprise, more than likely, procession is a function of the proton's decay, which DIRECTIONALLY releases MEQ up through the Watch Glass. All of which, brings us to a very interesting discussion of Gravity and Tilt?

Chapter 10: Gravity and Tilt

Even though the idea of atomic nuclei slowly rotating to align with other atomic nuclei, over millions of years in deep-dark-cold positive space seems a bit far fetch, there has to be a gestation period to the matter that forms planets, otherwise they'd just break into asteroids? Or in other words, a seed of similarly aligned elements, a core if you may, guides the planet's gestation period? All of which, will bring us to gravity and tilt...hopefully.

As my theory goes, a star's last degradation produces tons of protons, which results in the "fast" diffusion of energy away from the old star, i.e. an explosion. But, and this is important, this controlled explosion can ALSO form a cocoon around a mass of ISOLATED protons, which can slowly, and relatively COLDLY, undergo fusion to form heavier elements. All of which, brings us to another ultimate question: Does the SPIN of atoms facilitate the alignment of adjacent atoms? Unfortunately, to answer this, we have to ask a very uncomfortable question: Do red PMEQ traverse the atomic nucleus in search of blue NMEQ?

So what do we know about this question? Well, we know that the movement of negative electrons in a circle can focus Earth's relatively positive MEQ, which is the basis of NMR technology. (Now just imagine that on the atomic scale, where electrons/NMEQ focus PMEQ through the atomic nucleus?)

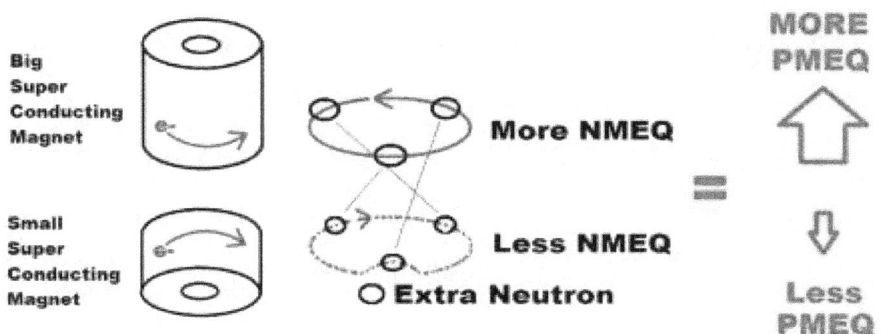

Figure 33: The Question?

With this in mind, how and why did the elements in Earth's core slowly align? Well, let's start from the end and work backwards.

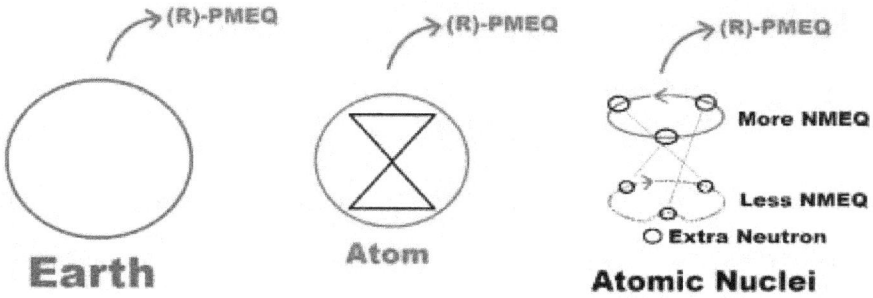

Figure 34: Patterns...to tilt?

As you can surmise, the excessive amount of the "More NMEQ" on top focuses MORE (R)-PMEQ out the top of the atomic nucleus, which results in the TOP of the atomic nuclei "leaning" towards a negatively diffusive entity. Or in Earth terms, MORE relatively positive (R)-PMEQ out of Earth's top will cause Earth to tilt towards the negatively diffusive Sun. All of which, brings us to the alignment of the atoms in Earth's core.

For a second, let's imagine that Earth's core is comprised of **dense** SPIN-ALIGNED-Tellurium. Or in other words, the traditional view of chemical bonds is NOT applicable under these pressure conditions, which allows the Tellurium to align based upon the atomic SPIN.

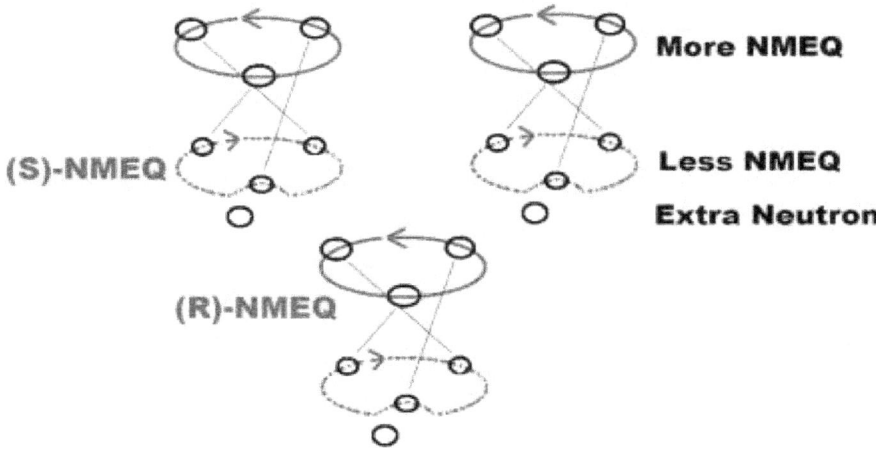

Figure 35: The WONDERS of Super Dense Matter?

As you can imagine, as a result of the insanely dense conditions about the Earth's core, there are LESS electrons "holding" the atoms apart. And as such, the attraction/repulsion between (R) & (S) NMEQ is the

FORCE that results in the core's alignment over millions of years...until this "mass" happens upon a negatively diffusive star. All of which, results in Isotopic Relaxation/Degradation to create a layer of solid rock, then magma, and finally, a cold Tellurium core. (BTW, let me reiterate a VERY important postulate: The relativity positive PMEQ being released from the core PUSHES the negative heat, i.e. TEQ from the magma layer, AWAY from the core, which is how the core stays cool?) And now...what we've all been waiting for, and secretly dreading, MAGNETIC INVERSION!

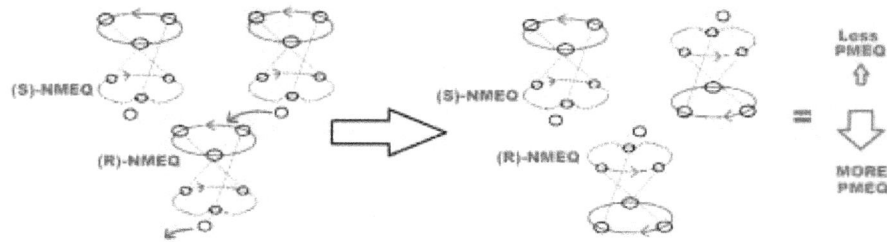

Figure 36: Magnetic Inversion

As you can see, the mass movement of the extra neutrons results in a change of "PMEQ focusing" power, which will cause a magnetic inversion. Also, if you're REALLY paying close attention to the figure, you'll surmise that MOST magnetic inversions are the result of a percentage of extra neutron movement, which results in a SPECTRUM of possible magnetic inversions? And finally, I prefer to use the term "atomic loci" to describe the atomic nuclei in planetary cores, because they're in such a unique environment?

With all this in mind, I hope that someone will realize that NOT EVERY planetary core is the same because the FUNDIMENTAL focusing power of the MEQ being released by the decay of electrons is based upon the "strong nuclear force" within the atomic loci, i.e. the NMEQ surging around the tops and bottoms of the Neuproz Groupings. And hopefully, with some careful statistical calculations, a scientist could correlate the abundance of the elements in the planet, in association with the tilt, to determine the average composition of the planetary core, which would help in the VARIABLE gravity calculation? In any event, here's an odd experiment that might produce some data to support my idea of a **dense** SPIN-ALIGNED-Tellurium core.

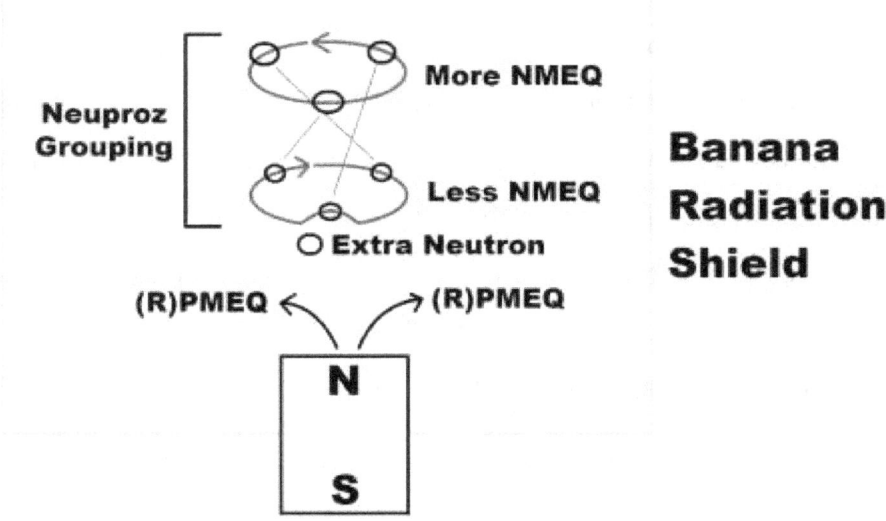

Figure 37: Wild Idea?

In theory, if (R&S) PMEQ are traversing the atomic nucleus based upon their attraction to the (R&S) NMEQ, which are racing around the tops/bottoms of the Neuproz Groupings, then the inundation of the atom with ONLY one type of PMEQ, i.e. the North PMEQ, should noticeably change the atomic orbital characteristics. As to why? Well, that has to do with the complex overlapping of energy flow as it relates to electron movement. Or in simpler terms, just ONE type of PMEQ about the atomic nucleus will cause the atomic nucleus to expand. Actually, let me try and explain this a little better.

Opposite PMEQ are attracted to each other and engage in a slow decay/death swirl, which is anti-entropic in nature...until they obliterate to form plasma. Therefore, the relative positivity of this swirling couple is SMALLER in comparison to the equivalent number of JUST ONE TYPE of PMEQ. Or in other words, JUST ONE TYPE of PMEQ in the atomic nucleus will display more POSITIVITY, which will cause the atomic nucleus to expand, change the movement of the electrons, and modulate the element's characteristics. All of which, can be correlated back to the behavior of the **dense** SPIN-ALIGNED-Tellurium core...and how it inverts from time to time?

In conclusion, not every planetary core is the same. BUTT, the basis of magnetism under these dense conditions is FUNDIMENTALLY the same, i.e. the alignment of atoms based upon the attraction and repulsion of NMEQ, which are surging around the top and bottoms of Neuproz

Groupings. And finally, there's a spectrum of magnetic inversions based upon the composition and order of the atoms within the planetary core.

Chapter 11: Splitting Centripetal Acceleration

With the last chapter's slight knowledge-expansion with regards to the order/magnetism of planetary cores, hopefully, this is the perfect time to talk shit about the "universal" gravity constant...again. All of which, should enable me to split centripetal acceleration into two parts. But first, let's talk about trust.

The major problem with science is that it can't explain everything. For example, if you go outside, attach a heavy object to a rope, swing the rope around in a circle above your head, and let go, what happens? Well, your "current" physics teachers would have you believe that centripetal acceleration is "real," even though the heavy object you were swinging around on a rope did NOT accelerate towards your head. (BTW, please don't do this experiment when other people's heads are in a close proximity!) All of which, brings us to the point of this chapter: Splitting Centripetal Acceleration into two parts...based upon magnetism.

Part One: Magnetic Centripetal Acceleration

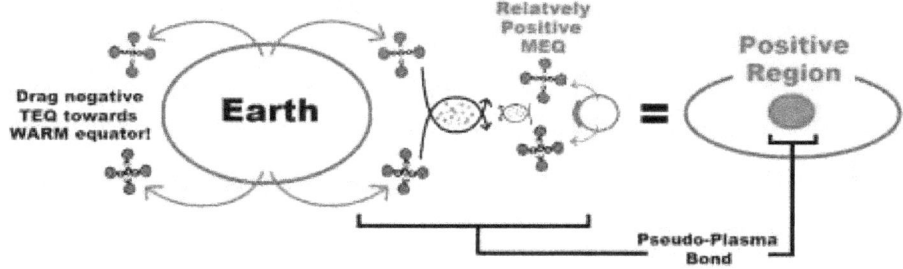

Figure 38: Magnetic Centripetal Acceleration

As you probably **don't** want to surmise, since there's a lot going on in this figure, the Earth's HUGE magnetosphere "complicates" the slow swirl decay of the Moon's PMEQ, which increases the positivity on the side of the Moon that is facing the Earth, i.e. the exaggerated red dot on the right side of the figure. All of which, causes the negative elements in the Moon to "accelerate" towards this red/positive region between the Earth and the Moon.

Part Two: Electronic Centripetal Acceleration

Figure 39: A Spinning NMR

As you can see, the ENHANCED collision of Earth's relatively positive MEQ into the sides of the SPINNING superconducting magnet will result in a red positive plasma layer, which the electrons will **move** towards. All of which, is in OPPOSITE direction of Magnetic Centripetal Acceleration.

In conclusion, centripetal acceleration is relative to the magnetic environment. Or in simpler terms, centripetal acceleration is a bunch of whoa used to explain data without the understanding of Quanta Dynamics, which is somehow related back to the concept of a "universal" gravity constant…even though magnetism exists as a spectrum? And finally, hopefully, this explanation results in more "trust" towards science, so we can design spaceships with more thrust?

Chapter 12: Propulsion & Fusion

Recently, I suggested that taking a core sample of the Moon, since it doesn't have a magma layer, would enable future scientists to make better celestial postulates. And in addition to that, scientists could purify "similar" protons on the Moon to make tritium, which could facilitate interstellar travel?

We know that Moon has a very weak magnetosphere AND protons are made of magnetism, since they continually release magnetism over billions of years. Therefore, as I've previously discussed, less distortion of the proton's electronics by the Moon's weak magnetosphere should allow for the purification of "similar" watch-glass protons. And finally, "similar" electronic structures result in the "sharing" of energy, i.e. strong and weak nuclear forces. All of which, makes me wonder: Why do the current scientists think that adding in more negative energy, i.e. heat, is the ONLY way to accomplish fusion? Or in simpler terms, short-term radioisotopes are made in super-colliders, which are really-REALLY cold? Any who, with just a little pressure about "similar" protons, their electronic watch-glass structures should align, i.e. allow for the sharing of NMEQ.

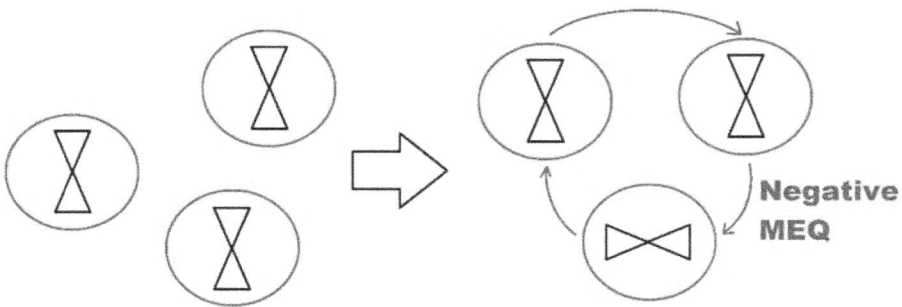

Figure 40: Tritium Fusion

All of which, could be used to make an intergalactic thruster.

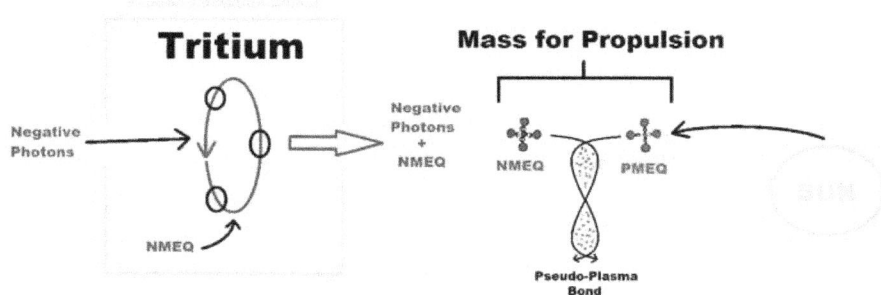

Figure 41: NMEQ Propulsion

As you can see, since light is negative, it should "knock" some NMEQ out of tritium, which reacts violently with the PMEQ, released by stars, to form plasma, i.e. thrust. All of which, brings us to this idea: Stars are "electronically" interacting with each other, i.e. Magnetic Super Highways.

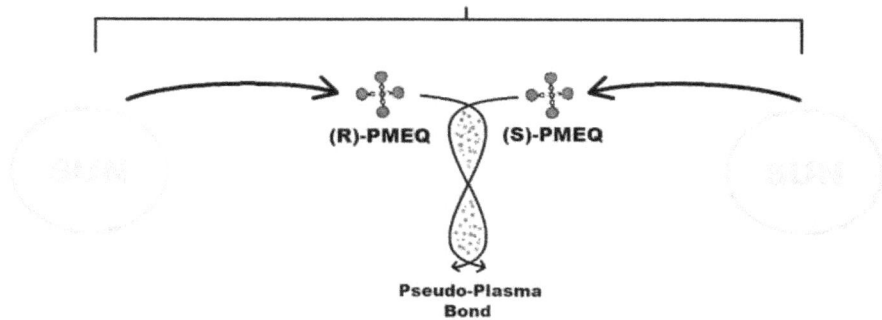

Figure 42: Magnetic Super Highways

Stars are really far apart, that's a no brainer. But, galaxies are held TOGETHER by electronic interactions between stars, which means that there's energy between the stars. Unfortunately, as of right now, we do NOT associate plasma with HEAT, which means we have no way of detecting these extra dense regions of plasma, i.e. Magnetic Super Highways. (Maybe comparison angular diffraction of light? Or in simpler terms, multiple observation points that detect odd refraction areas in deep-dark-cold positive space. But, that would mean we'd have to stop fighting and launch more space expeditions?)

In conclusion, the inter-electronics of atomic particles, i.e. NMEQ, which are causing elements to have different spins, can create propulsion in the area between stars, since stars are electronically interacting, i.e. Magnetic Super Highways. Now, all we have to do, is convince everyone to stop fighting and invest in space exploration? (FYI, more than likely, the purification of "similar" protons will only be accomplished on the Moon, which means whomever wants to own the technology of intergalactic space travel might want to own a biggest portion of the Moon?)

Chapter 13: Termites

So here's the thing about termites: They're sort of weird. Are they ants? Are they evolved ants? Why aren't termites a larger biomass if they've existed for so long? Well, here's my theory: Trees used to be really "weird" before the Ozone Event that caused the atmosphere to evolve, i.e. more carbon dioxide, which means termites did NOT exist on primordial Moon-Earth.

For a moment, let's imagine that primordial trees used oxygen as the final electron acceptor on Moon-Earth. If this is true, then eating a tree would be like eating an animal, i.e. the same type of molecules and tissue. Butt then, the Ozone Event changed the atmosphere, which caused the evolution of new Fungi and NEW trees. All of which means, termites didn't evolve until after the Ozone Event...sometime around Earth's new tilt? Or in other words, **termites** haven't been around very long, which is why they don't account for much biomass. Or in the simplest terms, ants have been around forever, which is why they're the second largest biomass.

In conclusion, this was a short chapter about a LONG process. Butt, does that make it any less logical? I mean, if our current "trees" have been around FOREVER, then wouldn't termites be the LARGEST biomass? (FYI, termites are kind of freaks in as much as there's lots of different types of ants that eat "leafy" vegetation, but only termites eat wood. Or in scientific terms, termites evolved sometime AFTER the dinosaur extinction…round about the time of the NEW trees?)

Chapter 14: Seeing & Communicating

If there were a highly advanced society trying to enlighten distant worlds, to their existence, they would probably want to do it VERY carefully. First, they'd probably would NOT use radio waves, because most Neanderthal societies will be able to detect it. Second, they'd would NOT use an SOS with regards to the light being released from their star. With those bad "ideas" out of the way, here's a better one: An undetectable signal to everyone that doesn't have the code to break it. In any event, here's how I'd send a non-detectable message to other like-minded beings.

With the knowledge of Quanta Dynamics, as it relates to unique photons with unique charges that ONLY respond to magnetic fields to produce wavelike patterns, I would encode a message in our star's light by creating an organized pattern of photons. Or in other words, I would separate unique photons, then re-organize them into an un-natural arrangement, which can ONLY be detected by the technologically of an advanced society. Therefore, the "signal" would just seem like light from a star to the uneducated Neanderthal worlds, but it would be a beacon of "ordered-hope" to any world that breaks through to a higher plain of consciousness, i.e. Quanta Dynamics?

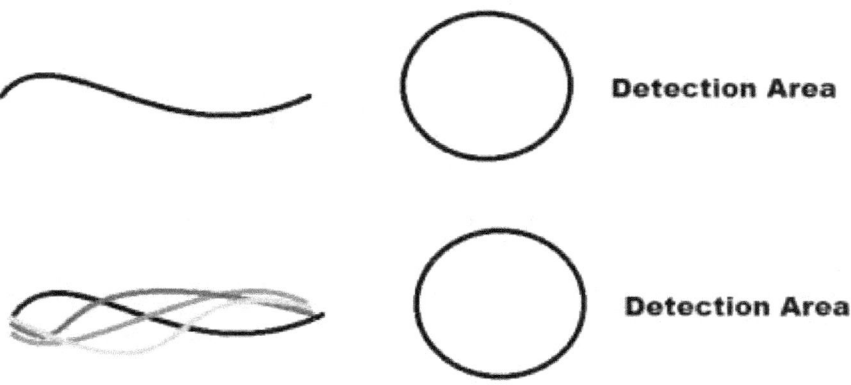

Figure 43: Abnormally Dense Light

As you can surmise, this data point will not mean anything until a second angle of the star is observed. If the second angle contains "normal" photon density, then the finding is of scientific significance. And the greater variance between these two detections will solidify the scientific finding? (BTW, most of the smart people are probably wondering the following: Ummm, how would the aliens know where to be focusing the "ordered" light? Well, here's my answer: Hey, if they're a super-cognizant society, then more than likely they've discovered a way to detect planets with life?)

In conclusion, even if there's a highly intelligent species encoding abnormal 'less'-entropic photon patterns within their star's light, it's highly unlikely that we'll detect this signal until we have a second observation point, i.e. a Pluto scientific base? In any event, another eclipse was today and I didn't take the time to go look at it…oh well? (FYI, eclipses were bad-omens, from God, because so many people went blind from looking at them?)

Chapter 15: Which way is up?

I don't know how the philosophy of our existence became tied to the relative directionality of living on Earth, but ONE thing is for sure: The universe probably has a specific "up and down," but we have no clue as how this relates to Earth's relative directionality.

With that in mind, "up and down" on Earth has been correlated to being free like a bird or being trapped by the warm Earth, which somehow got correlated to heaven and hell. Now, I'm **not** saying that the construct of heaven and hell has become a problem when addressing the "idea" of climate change, but I do think humanity has a tendency NOT to respect the things that are "below" them. Granted, the whole idea of "mother Earth" was supposed to facilitate some rebuke of the heaven/hell construct, BUTT…women still don't get the respect they deserve. All of which, correlates back to people NOT respecting Earth?

So where am I going with this? Well, I don't know…because I don't know which way is up, relatively speaking of course. What I do know is: We can do better, which starts with equal education reform? Simply divide the money for education EQUALLY across the state? We need more heads in the clouds to protect Earth, humanity, and America?

www.ingramcontent.com/pod-product-compliance
Lightning Source LLC
Chambersburg PA
CBHW071425220526
45469CB00004B/1436